Airliners at LAX
Los Angeles International Airport
1956 - 1976

Dedication

to Patricia and Victor

Acknowledgments

Thanks to Jon Proctor, for his contributions as editor
and help in finding obscure historical information.
The aircraft color photographs in this volume are from
the author's collection. The black-and-white photographs
are courtesy of the City of Los Angeles Department of Airports.

© 1997 by Robert D. Archer
All rights reserved.
ISBN 0-9626730-6-4

First Edition June 1997
Printed and bound in Hong Kong

No part of this publication may be reproduced, stored in a retrieval system, or transmitted by any means without first obtaining written permission of the publisher.

The publisher acknowledges that certain terms, logos, names, and model designations are property of the trademark holder. They are used here for correct identification purposes only.

Series Editor: Jon Proctor
Series Design: Randy Wilhelm, Keokee Company, Sandpoint, Idaho
Copy Editor: Fred Chan

Published by:

WORLD TRANSPORT PRESS, INC.

P. O. Box 521238
Miami, Fla. 33152-1238, USA
tel: +1 305 477 7163 • fax: +1 305 599 1995

AIRLINERS AT LAX
LOS ANGELES INTERNATIONAL AIRPORT
1956 - 1976

By Robert D. Archer

Introduction

Los Angeles Airport has grown tremendously in both traffic and facilities since 1956. That happened to be the year the author came to the Southern California area as an aeronautical engineer. The scene at LAX (as it is code-named) proved to be very exciting and accessible for photography, mainly due to the airport being surrounded on all sides by public highways. Also, the Santa Fe railroad tracks (rarely used during the day) were immediately adjacent to the fence at the east end of the runways.

This, and the generally good weather, allowed one to get quality photographs of aircraft landing and taking off, including all scheduled airline traffic, plus a modicum of light airplanes. There was also a lot of military aircraft doing business at the adjacent North American and Douglas aircraft plants.

Looking back over these years, it is now possible to see the various changes that took place. The all-omnipotent piston engine quickly gave way to the surge of new jet engines, plus a sprinkling of turboprops, beginning in 1959. The newest jet power plants were quickly followed by the more fuel-efficient turbofans, and there was also a shift in the balance of the commercial aircraft manufacturers. It is possible to trace the slow decline of Douglas and the rapid rise of Boeing, together with the disappearance of Convair and Martin, plus other lesser manufacturers, in the years covered by this book.

The author was fortunate to travel quite often in most of the aircraft types shown, taking him all over the United States, plus trips to Canada, Mexico, Europe, South America, and the South Pacific. The aircraft on these trips were flown by extremely professional crews and cabin staff, and the author would like to say thank you to all of them for their care and skills.

It is also fascinating to follow the changes in the liveries applied by the airlines to their aircraft, to make them stand out on crowded ramps. When the author was taking these pictures it was difficult to obtain the data now available to any interested person, and indeed, he was often asked "Why are you taking pictures of the aircraft?" The answer, of course, is that they are here today, but gone tomorrow!

The adjacent black and white photographs of the original 1950s facilities show the tremendous changes that have taken place, and also bring back many pleasant memories. It is hoped that this book will help those who were there to remember what it was like, and to show those who were unable to be present how the airliners at LAX have changed. Enjoy!

What better way to begin a nostalgic tour of the airliners serving LAX than with one of TWA's postwar Douglas DC-3s. In natural metal finish with red stripes, this aircraft typifies the simple, dignified paint schemes of a bygone era. This was one of the last few TWA DC-3s (note the number 2 at the top of the vertical tail) used for ILS and instrument pilot checkouts.

Built as a military C-47, N86544 was originally USAAF 42-68762. Bought by TWA on October 17, 1945, it later became one of a few kept for non-revenue purposes, eventually being sold in April 1957. It went down to Brazil with Cruzeiro as PP-CES, then with other carriers, prior to being broken up in May 1980.

Fourth from last of Transcontinental & Western Air's 049s, N90823 was delivered on March 28, 1947. Given the name Star of the Yellow Sea, *and Fleet Number 512, the Constellation stayed with the renamed Trans World Airlines until March 1962, then passed through eight owners before going to South America in July 1976. It was withdrawn from use, and broken up in 1980.*

TWA, as one of the "big four" regulated US airlines, became legendary for its worldwide use of the Lockheed Constellation family. Here we see an early model 049 on finals, showing off its large flaps and downward curving fuselage. This fuselage contour had been adopted to reduce nose landing gear length, but it remained as one of the longest used on a commercial transport.

Delta Air Lines did not serve LAX directly at this time, but was part of the American-Delta-National interchange service between LAX and Miami, with intermediate stops. Delta bought ten Douglas DC-7s, and this shows the fourth one on finals. It is finished in the attractive "Royal Crown" livery.

Delta took delivery of N4874C, Fleet Number 704, on April 10, 1954. It served with them until March 31, 1968, when it was stored at Ontario, California. Sold in 1971, it was broken up at Barstow, California, in 1976.

N37568 was handed over to United Air Lines as Mainliner Des Moines on July 29, 1953, being sold by them 15 years later. Going through several users, it was re-registered as N66DG by Pacific Alaska Airlines in 1972, and finally stored in 1984.

United was a staunch user of Douglas types, and one of its DC-6Bs is seen on finals, with a nice cloud background typical of January in southern California. The DC-6B was the best money-making commercial transport of the era, carrying 82 tourist (economy now) passengers at a 34-inch seat pitch in a five-abreast layout. It also turned out to be the most successful of the DC-4/DC-7 series with no less than 288 being built; the last airframe was delivered in 1958.

Trying to get ahead of Douglas Aircraft, Lockheed installed new Wright Turbo-Compound engines in the 1049 Constellation, resulting in the models 1049C, D, and E. An improved version of the 1049E was the 1049G, which could be fitted with the unusual tip tanks (probably from their P-80 and P2V experience) for long-range operations. It quickly became a TWA stalwart. Unfortunately, the engine proved to be very unreliable and was the eventual downfall of the 1049G and competing DC-7 series. This "Super G," as TWA called the type, is seen on a wet ramp awaiting its next trip. The author had an interesting New York– London trip on a 1049G which suffered two separate engine failures, resulting in unscheduled landings at both Gander and Shannon. At Shannon, seven other Connies and DC-7s were visible with one or more engines feathered.

TWA received this 1049G on March 14, 1955, and named it Star of Chambord, Fleet Number 105. California Airmotive Corp. bought it in December 1966. Withdrawn from use and stored at Fox Field, Lancaster, California, the Connie was broken up in 1971.

Delivered in December 1944, USAAF C-54, 42-72382, was converted to DC-4 standards by Douglas, and sold to United as NC30043, Mainliner Golden Gate, on March 14, 1946. Going to Brazil in 1957 as PP-LES, it was stored in 1970, and scrapped in the mid-1970s.

Another Douglas product in United colors shows up on finals, this time a venerable DC-4. The type began the modern Douglas four-engine transport family, being a completely new design, rather than a slimmed-down version of the original DC-4E prototype of 1938. Ordered before World War II, it did not reach delivery until early 1942, as C-54s to the Army Air Forces. First use of the civil DC-4 was in mid-1946 with American Airlines. Unpressurized, it normally seated 44 passengers. United's original color scheme shows up well against a beautiful sky.

A new visitor in summer of 1956 was the Bristol Britannia 102, G-ANBJ, (the tenth built) on a sales tour of the Americas. The British aircraft industry was trying to capitalize on its pioneering turboprop and jet aircraft (the Viscount, Comet, and Britannia), and hoped for sales to the US airlines. Very sleek looking, the Britannia was the first turboprop aircraft to be used on North Atlantic service. The author had just arrived in California after spending two years on the re-design of the Britannia as the CL-28 for the Canadian Air Force.

Bristol Britannia 100, G-ANBJ, was delivered to British Overseas Airways Corporation on November 22, 1956, and served with the airline until November 1962, having been leased to Malayan Airways twice in 1962. Britannia Airways used it between April 1965 and July 1970. It was broken up in February 1971.

The Britannia was powered by four Bristol Proteus turboprops of 3,900 hp each and proved to be a very comfortable, smooth aircraft in service. The author flew in a BOAC example from Miami to Jamaica and return in 1960. Although Capital and Northeast Airlines were both interested in buying the type, the FAA raised concerns about the main landing gear bogy not castoring. This would have caused major re-design of a complex, expensive system and sales fell through (this concern hurt Douglas financially when first designing the DC-8 main gear, but it was quietly forgotten when the Boeing 707 entered service).

Mentioned earlier was the Wright Turbo-compound engine in the 1049G. The Douglas answer was the DC-7 series with the same power plant, but it proved to be just as unreliable, and the aircraft were much noisier and suffered from greater vibration levels. Though not as popular as their predecessors, 338 of the DC-7/-7B/-7C family were produced. We see a United DC-7 on the ramp in the later paint scheme, being serviced for its next trip. This version gave the airlines nonstop transcontinental capability both east and westbound (though TWA 1049s actually operated the first eastbound nonstop service). The author vividly remembers on several night flights, the long flames of the engine exhausts licking over the top of wings filled with 130 octane fuel.

Mainliner Hartford was delivered on January 15, 1957. Renamed Mainliner Columbus *before being sold in July 1963, it was stored and later broken up.*

DC-6B N90961 joined Continental Airlines on January 5, 1955. Leased to Air Micronesia between 1968 and 1976, it was converted to a dual configuration and carried main-deck cargo as well as passengers. Sold to Aerospace Products, the propliner ended up being impounded at Bogota, Colombia, in August 1979.

One of the medium-size carriers serving LAX was Continental Airlines; here is one of its DC-6Bs just starting its takeoff roll on Runway 25 Right. This particular aircraft was used on a joint Continental/ United service, hence the dual markings. The early, low control tower and the background Hollywood Hills show up well on a beautiful spring day in 1957. Note the TWA hanger and Connies to the right of the DC-6B.

My, what do we have here? One of only three TWA Douglas C-54Bs is seen late in the day after being washed. Delivered in 1946, they were originally used in passenger service, but relegated mainly to freighter use when the Lockheed 049s arrived. The type were used to haul engines and other large items from the airlines' Kansas City base, and rarely to ferry personnel. It carries a very simple paint scheme like the DC-3s, without a white top, and has the "Sky Freighter" inscription below the cockpit.

Built in 1945, USAAF C-54B-1-DC, 42-72436, was acquired by TWA as NC34577 in 1946. Named Le Moulmein Pagoda, *Fleet Number 607 went to Eastern Aircraft Sales in March 1958. Ending up in Nicaragua as AN-AMK, it was stored, then broken up in 1967.*

N8408H was Western's eighth example, delivered on October 5, 1948. Sold in 1961 to Han Jin Transportation Company, it became HL-25 with Air Korea, and later flew with Japan Domestic Airlines as JA5092. The Convair-Liner returned to the US in 1969, again as N8408H, and in 1976 was purchased by Texas Instruments. Withdrawn from use and stored at Chino, California, it is now preserved at the Planes of Fame Museum at Chino.

A major user of LAX was Western Airlines. One of its Convair 240s is seen on the ramp, with two United 340s far behind on the left. Western used these aircraft on services to San Diego, Palm Springs, Las Vegas, and other local towns. Note the aft stairway on the aircraft and the minimal ground-handling equipment needed. The color scheme is Western's second postwar example, and still reflects the use of "Air Lines" in the carrier's name, even though "Airlines" became the norm from April 1954.

A Western DC-6B is seen taxiing for departure to Seattle on a spring day in 1957. Western used these, its largest aircraft, mainly on the runs to Seattle/Tacoma, Salt Lake City, and Minneapolis/St. Paul. Very comfortable to fly in, Western's had a six-seat lounge at the rear of the fuselage, with a large coffee table in the middle. At Christmastime it was laden with fruit, nuts, and candy, and ever-attentive stewardesses ensured a never ending flow of the best California champagne. Yes, the author enjoyed this on three trips.

Western Airlines acquired N93116 new, on August 15, 1956, and sold it to Japan Air Lines as JA6208 on February 27, 1962. Passing through seven registrations in four countries, it ended up as N120C with Integrated Marketing Management in August 1992.

USAAF C-54B-15-DO, 43-17183, was bought by Braniff Airways as NC86573 in November 1945. Western acquired it in March 1954. In May 1957, the aircraft was sold to Loide Aereo National of Brazil, and re-registered as PP-LEL. It was broken up in 1969.

Western's first postwar four-engine type was the DC-4; one is seen in its original color scheme, with a later DC-6B in the newer livery behind. Contrary to first impressions, the round porthole windows were used on the unpressurized DC-4s, while the rectangular windows were used on the pressurized DC-6s and later types. Note the difference in the stairways being used at both aircraft, with the earlier, more austere one, situated at the DC-4.

Another hopeful foreign visitor to LAX was the Sud Caravelle, in mid-1957. F-BHHI was the second prototype, making its first flight only a year earlier. The French aircraft was the first short-haul jet airliner, the first twin-engined example, and the first with aft-mounted engines, Rolls-Royce Avons in this case. Using the nose and cockpit of the British de Havilland Comet, the sleek lines of the Caravelle are evident here. Note the unusual triangular-shaped windows.

F-BHHI first flew as F-WHHI on May 6, 1956, and was re-registered on March 19, 1957. Withdrawn from use by Sud-Aviation in 1968, it went to the Centre d'instruction Vilgenis. The Caravelle's nose was placed in the Musee de l'Air at Le Bourget in 1976.

The very clean installation of the Avon engines and the mid-set horizontal tail are seen in this view of the Caravelle. Also visible are the two large fences on the moderately swept wing. The legend "SE-210-2" on the rudder is the usual French method of designating their aircraft. Unlike the Britannia, the Caravelle was put into service within the United States by United Air Lines. Twenty were purchased for use on routes east of the Rocky Mountains. The unique position of the horizontal tail ensured that the Caravelle never suffered from deep-stall loss of control at high angles of attack.

Beside the regular airlines serving LAX there were many visits by non-scheduled and freight aircraft. Here is a very clean Austral Curtiss C-46 from Argentina, parked at the eastward end of the ramp at LAX, next to Aviation Boulevard. It carries a penguin insignia on the vertical tail.

C-46A-45-CU started with the USAAF as 42-96661 in August 1944, and was turned over to the US Navy the same day. Bought by L. B. Smith Aircraft Corp. in June 1956 as N3943A, it went 18 months later to Austral in Argentina as LV-FSA, named *Caiquen*. The aircraft returned to the United States as N8040Y in July 1979 and was stored.

N7303C, Star of Vermont, was delivered on May 26, 1957. Serving with TWA for less than six years, it was withdrawn from use and stored at Kansas City in December 1962, being broken up for spares in February 1965.

Continually vying for increased sales with Douglas, Lockheed produced the ultimate Constellation variant in the form of the 1649A. This had a totally new wing, more than 27 feet longer than the earlier 1049 series. This met the need for more fuel capacity to ensure year-round nonstop trans-Atlantic flights in both directions. TWA was the biggest user of the type, which it referred to as the "Jetstream." One is seen on final approach, only four weeks after entering service.

This dead head-on view of a TWA 1649A on approach shows very clearly the extent of the increased wing span. To decrease the cabin noise and vibration levels, Lockheed also moved the engines more than five feet outboard, and fitted them with larger diameter, slower rotating propellers. This gave the 1649A the greatest width between the main landing gears of any transport then flying, and required careful taxiing. The 1649A was an extremely comfortable aircraft to travel in, the author being on one London–New York flight that lasted more than 14 hours due to bad weather over the East Coast.

N7311C, Star of Ebro, was handed over to the carrier on May 31, 1957. It underwent a freighter conversion in October 1960, and was sold to California Airmotive Corp. in September 1967. Trans American Leasing bought the Starliner in December 1968; its end came in March 1969 when a forced landing in Chile damaged it beyond repair.

TWA's twentieth 1649A arrived new on July 26, 1957. Following barely four years of service, it was sold to Trans Atlantica of Argentina as LV-PHV, and re-registered as LV-HCD shortly after delivery in August 1961. It was reclaimed by TWA in June 1964, long after Trans Atlantica ceased operations. Sold to a company in Florida, it was withdrawn from use and stored in Fort Lauderdale by May 1966. Sold again in January 1969, the Starliner was broken up in May 1970.

A left side close-up of another TWA 1649A landing shows its very clean finish. Although intended to compete with the DC-7C, it did not enter service until a year later, and did not last very long in service because BOAC's faster Britannias were in use over the North Atlantic only six months later, and 707s, Comet 4s, and DC-8s barely eleven months after that. The story of the 1649A is that of an aircraft too late for its intended traffic, but it was the best piston-engine type during its short life.

By mid-1957, the DC-3 was out of mainstream airline use at LAX, but survived for many more years with small local carriers serving special niches. One of these was Catalina Pacific Airlines, which flew to Catalina Island, a 27-mile ocean flight. Here is one of its DC-3s in a distinctive green and red finish about to touch down at LAX.

This C-47B started with the USAAF as 43-49414 in November 1944, and saw war service in Europe. It went through several owners in Australia before returning stateside in December 1956 for use by Catalina Pacific Airlines as N55L. The aircraft was leased to Avalon Air Transport, 1962-1963. The last known operator was Continental Air Service, Los Angeles, between 1966 and 1971.

N6334C, United's Mainliner Baltimore, *joined the carrier on February 28, 1957. Sold to Charlotte Aircraft Leasing on September 17, 1962, it was stored pending resale. This never occurred and the DC-7 was broken up.*

United Air Lines stuck to the original DC-7 model with a fleet of 53, and never bought any DC-7B or DC-7C variants. Here we see one on the by-then small ramp at LAX, with two of the airline's very smartly dressed stewardesses about to board their aircraft. Despite the increase in the nominal number of seats from the 44 on the DC-4 to the 92 tourist seats on the DC-7, the ground-handling equipment remained essentially unchanged.

One of the more unusual types operating into LAX was the Martin 202, flown by Southwest Airways (no connection with today's Southwest Airlines). Competitor to the Convair 240, the 202 suffered from some early major structural problems, and never recovered ground. In the background is a DC-3 of Southwest, with an earlier version of the company logo on the rear fuselage.

N93056 was delivered to Northwest Orient Airlines on May 29, 1948, and flew with them until April 1, 1952. After serving with Pioneer Air Lines, it was bought by Southwest (later renamed Pacific Air Lines), in March 1958, then sold to TWA two years later in partial exchange for Martin 404s. Following another decade in various roles, it was broken up at Sarasota, Florida.

A surplus C-47 acquired by Southwest in 1946, N63105 was sold in 1958 by Pacific Air Lines to Banfe Aviation. It eventually migrated to Southeast Asia and was last reported flying with Air Vietnam in 1970.

United's Mainliner Salinas *began its career in October 14, 1952. It was sold to Turbo-Prop Conversions on December 6, 1960, then to Frontier Airlines on July 20, 1962. Converted to a Convair 580 in July 1965, Frontier operated the propliner for two years before selling it to Sierra Pacific Airlines on October 15, 1974. It then went through several leases until withdrawn from use and stored at Marana, Arizona, in 1992.*

We have referred to United Air Lines Convair 340s earlier and here is one on finals, showing its clean lines. The 340 was a lengthened version of the 240 and gained a little extra speed from combining the engine exhausts into two large ones above the wing. This gave a modicum of extra thrust, but also made the aircraft very noisy for passengers sitting behind the wing, contrary to most piston engine types. The author flew in one up the East Coast in rough weather and found the turbulence and noise very conducive to motion discomfort, especially when watching most of his fellow travelers using their airsick bags.

Continental Airlines continued to grow in the late 1950s due to excellent service, and bought five of the long-range Douglas DC-7Bs for routes to the east. Here one is seen taxiing in the summer of 1958. The considerably longer fuselage of the DC-7 versus the DC-4 clearly shows up.

Continental took delivery of DC-7B, N8211H, on April 5, 1957, naming it City of Los Angeles. *In March 1964, '11H went to Ecuador as HC-AIP. It came to a sticky end, being damaged beyond repair after taxiing into a ditch at Miami, Florida, in March 1966.*

This DC-3A-191 is a genuine prewar aircraft, being handed over to United on January 5, 1937, as NC16063. It passed to the Babb Co., December 18, 1956, and was operated as N498 by Bonanza until 1961, then ended up with Hanger 10, Inc. as N515AC, at McAllen, Texas, its last known disposition, in 1984.

Yet another small West Coast carrier was Bonanza Air Lines, which started service between LAX and Las Vegas in 1946, with the ubiquitous DC-3. One is seen just prior to landing at LAX. Bonanza later expanded to serve Torrance, Apple Valley and other southern California towns.

Ultimate stretch of the Douglas DC-7 family was the DC-7C, commonly called the "Seven Seas." 120 of these were produced (the most of the three DC-7 versions) and this is a Pan American example on finals. The extended wing center-section added the same benefits to the DC-7C as the new wing on the 1649A Starliner, but at a considerably lower cost. It resulted in the engines moving outboard, giving the main landing gear a much-increased track (though still slightly less than the 1649A), as clearly evidenced here. One unseen result of the DC-7's successful sales was to make Douglas very reluctant to spend the vast amount of money to go ahead with a new jet-powered transport, giving a very aggressive Boeing a head-start with its Model 367-80 prototype (which flew into LAX on a very short visit, on a November 1956 night).

N739PA joined Pan Am on June 22, 1956, as Clipper Flora Temple. *It was converted to a freighter just four years later, and re-registered as N7398A. Going through a number of operators, including as G-AVXH in 1967, it was stored at Miami, Florida, in November 1973, and later broken up.*

This C-47A-70-DL went to the USAAF as 42-100757, on December 8, 1943, and served with the 8th Air Force in England until August 8, 1945. Sold to Southwest in 1946 as NC54370, it later became N54370. Pacific Air Lines sold the trusty Douglas to Amering Turkey Breeding Farms in March 1964. It crashed just over three years later, on March 24, 1967, at Merced, California.

In March 1958, Southwest Airways had been renamed Pacific Air Lines, but was still flying DC-3s. Seen on finals, these by-now small transports seemed to scurry in and out, as if in a real hurry to get out of the way of the major airlines' bigger aircraft. In time, many of these small startup carriers were to merge into larger, but still medium-size, airlines, as will be seen later.

By 1959, Bonanza had grown to the point at which it could afford to add the new Fairchild F27A "Silver Dart" (as the airline named them) turboprop aircraft to its fleet. Actually the Dutch Fokker F27 built under license in the United States, the type was very successful and greatly improved flight times. The author flew in one out of Las Vegas and found it to be a very comfortable, smooth and quiet type. However, the wing area required for good performance resulted in considerable bouncing around in the afternoon turbulence characteristic of the desert regions.

Bonanza Air Lines took delivery of N147L on February 26, 1959, and changed its registry to N747L five years later. Bonanza, Pacific, and West Coast merged to form Air West0 on March 7, 1968. The aircraft was upgraded to -F standards for Hughes Airwest on April 1, 1970. Leased out, it was damaged beyond repair at Jeddah, Saudi Arabia, on March 3, 1978, when the landing gear was inadvertently raised while the aircraft was parked at the gate.

TWA's N90831, Star of Switzerland, began life as a USAAF C-69, 42-94549, delivered on April 28, 1945. Turned over to the airline on October 1, 1948, Fleet Number 517 flew overseas and later domestic routes until April 1961. After passing through a dozen owners, it was acquired by the Pima Air Museum in April 1971, and restored to original TWA colors for static display.

Turboprops or not, TWA's Lockheed 049s continued to serve LAX on a regular basis, and here is one just above the final approach lights. The Connies were beginning to look a bit dated by mid-1959, as is underscored by the next picture....

Earlier, we had alluded to the Boeing Company's aggressive development of a new jet transport. Here is the result, an American Airlines 707-123 with the red carpet rolled out for its passengers. There is no doubt that this type changed the entire world airline industry in many ways never foreseen; indeed the effects are still happening, as witness the surprise Boeing-McDonnell Douglas merger announced in December 1996. Behind the stairway can be seen a United Air Lines DC-7; United would have to wait a little longer for its DC-8s to enter service. The jets made rapid travel to any part of the world commonplace, and propelled Los Angeles into an Asian trading powerhouse.

Flagship Maryland was American's seventh 707-123, delivered February 27, 1959. It was converted to a -123B in April 1961, and continued with AA until May 1977, when it was sold as N707AR to Atlantic Richfield, May 1977. Passing through three other owners, the 707 was last heard of at Houston, Texas, in 1990.

N732TW was the first 707 delivered to TWA and flew the company's inaugural jet service March 20, 1959, from San Francisco to New York. Tel Aviv-based Israel Aircraft Industries purchased Fleet Number 7732, along with several other non-fan powered -131s, at the end of 1971. Following refurbishment, the Boeing was sold to Trans European Airways and re-registered OO-TEC. It was withdrawn from service and stored at Brussels, Belgium, in 1982, and later broken up.

TWA was not far behind with 707 service from LAX; here one of its 707-131s is seen wearing a striking new color scheme, in contrast to the relatively unchanged American Airlines livery. This 707 has just turned onto the runway to line up for takeoff (note the nose gear angle). Full throttle and 4,000 gallons of distilled water will provide the necessary thrust to lift the reluctant aircraft "Up, up and away," as a popular saying then claimed. "Water buffaloes" was another name for these early 707s.

As mentioned earlier, Delta Air Lines aircraft first began appearing at LAX as part of the American-Delta-National interchange service to Florida. Delta bought only ten DC-7s, and then ten DC-7Bs, while awaiting its new DC-8s. Unable to compete with the jets on long flights, the DC-7s and 1049s were bumped down to shorter hauls, such as Dallas to Montgomery and Atlanta.

DC-7 N4878C spent its entire flying career with Delta, from January 25, 1956, until March 1968. Withdrawn from use, it was sold to BMR Aviation and stored at Ontario, California, where it was broken up in 1974.

Continental accepted N70775 on July 16, 1959. It had a short life, being destroyed in flight by a bomb over Centerville, Iowa, on May 22, 1962.

By 1959, Continental Airlines had grown into a major competitor, initiating "Golden Jet" service to Chicago, Kansas City, and Denver with a fleet of 707-124s. Like TWA, Continental applied a striking new finish to the big jets, including a mostly gold-finished vertical tail. This did not make any difference to the horrendous takeoff noise, but the new stewardess uniforms presented a striking impression, and combined with excellent service, made it a very popular carrier. Only the unfortunate locals heard the noise....

United Air Lines received its first Douglas DC-8s only a short while after the 707s entered service, and here the twentieth one built is seen on finals just 23 days after delivery. A handsome aircraft, the early DC-8 model suffered from some performance shortfalls, and competing airline crews liked to call the type the "DC-Late." Engineering re-design of the wing leading edge cured the problems and the term was quickly forgotten! Moreover, the DC-8 proved to be a very stretchable type; the later DC-8-60 and -70 series are still in major use as rapid freight overnight haulers, almost 40 years after the DC-8's first flight.

N8012U, a DC-8-11, was delivered to United on December 18, 1959, named Mainliner J. A. Herlihy. Upgraded to DC-8-21 on August 21, 1965, it went to Boeing on November 20, 1977, most probably in exchange for new aircraft on order. Later converted to a freighter, it went through a number of operators until being withdrawn from use in 1985, and broken up at Marana, Arizona.

Southwest Airways acquired N93047 from California Central on March 25, 1955. After being sold to TWA in June 1960, the 202 passed through two other owners, then went to RAPSA Panama as HP-398, in October 1964. It was broken up at Panama, circa 1970.

Having to hurry even more with the increasing jet traffic, the local Martin 202s still continued to serve LAX, and here is one of Pacific Air Lines' seen in January 1960 on final approach. Due to the greater turbulence behind the new jets, more separation between aircraft was now required, and it was obvious that a new, much larger, airport was needed. In fact, planning for this had been under way for some time and the expanded facility was only some 18 months away from being opened.

Like United, Western Airlines was still using its fleet of Douglas piston-engine types. A company DC-6B is seen approaching the airport, against a beautiful winter sky.

N93131, among the last of the DC-6Bs to be built, was delivered August 29, 1958, to Western. It became CC-CDO with LAN Chile on August 17, 1965, and was bought by the Chilean Air Force in 1973, as FAC 990. Withdrawn from use, the -6B was stored in 1980.

N7136C joined Western on July 10, 1959, less than a year after the carrier's final DC-6B delivery. Sold to Concare Aircraft Leasing November 27, 1970, the Electra passed through many operators, and was converted to a passenger/freighter configuration just prior to its sale to the Argentine Navy, in August 1983.

Western was, however, also taking delivery of a new fleet of 12 Lockheed 188A Electra turboprops during 1959. The second example is seen against the same picturesque sky. By now, two Electras belonging to other airlines had been lost in accidents, but the type was popular with the public. The author had worked on the design of the Electra empennage (as a subcontract to Northrop Aircraft), and flew to Seattle on one of Western's 188s in December 1959. Performance was good, but the aircraft suffered from the worst engine-propeller-airframe vibrations of any aircraft he had ever flown on as a passenger. Subsequent events bore this out.

The British aircraft industry had not been very successful in selling its pioneering turboprop and jet types to US airlines, but eventually Vickers was able to place 14 Viscount Type 812s with Continental Airlines. The fifth one is seen on finals. Continental initially used these aircraft on the route to Kansas City, and they proved to be very popular with passengers. However, the incredibly high-pitched screech from the tips of the Rolls-Royce Dart engines' centrifugal compressors was not at all popular with airport personnel.

Continental accepted N244V on July 8, 1958, and operated the Viscount 812 for eight years before selling it to Channel Airways, England, where it became G-ATUE. Withdrawn from use in May 1972, the turboprop was stored at East Midlands, England, and broken up in January 1974.

Only the 32nd Boeing 707 built, N739TW was among those sold to Israel Aircraft Industries. Leased briefly to Air Siam as HS-VGA, and Trans-European as OO-TEE, it was withdrawn from use in 1976 and stored, eventually being broken up.

At the end of the same day, a TWA 707-131 is seen pouring on the power, complete with water injection and much black smoke, prior to starting its takeoff run. Note the large fences erected in an attempt to divert the jet blast from traffic on Aviation Boulevard.

One week later, a PSA Lockheed 188C Electra is seen head-on, showing the short wing span and the four very large propellers. The Allison 501D engines were unusual in that they ran at only two speeds, and prop pitch was used to regulate aircraft speed. Note the smoke behind Number One engine.

N172PS had a long and checkered history, originally being delivered on November 30, 1959. It left PSA ten years later, and passed through many owners and lessors before being converted to a freighter in 1976. Following a tour in South America as TI-LRN with LACSA, it returned to the US, again as N172PS, in July 1977. Re-registered the following year as N340HA for use by Hawaiian Airlines, the Electra migrated to Zantop International Airlines in September 1980, where it remains active.

It was rare to see an aircraft landing at LAX with a propeller feathered, but this unusual event was caught on a United Air Lines DC-7. The Number Four engine was shut down (yes, it was one of those troublesome Wright Turbo-Compounds referred to earlier). This was not a major problem, just as long as the other engine on the same side did not decide to quit.

Until 1962, the only major visiting airline to LAX from south of the border was Mexicana, known during this period as CMA (Cia. Mexicana de Aviacion). One of its DC-6s is seen landing on Runway 25 Right. Mexicana is the fourth oldest airline in the world, tracing its start back to CMTA, founded in 1921.

Delivered new on December 8, 1950, XA-JOT spent 22 years with Mexicana. It was sold to Transportes Aereos de Cargo in January 1972, then removed from service and stored in 1980.

The Mexicana DC-6 is seen being refueled by a brilliant red Texaco truck on the ramp at LAX. Note the Chevrolet autos in the background, real collector items!

Another PSA Electra is seen just touching down on Runway 25 Right, with the Douglas El Segundo plant in the background. This was the airline's third Electra, in its original dark red color scheme. A really dazzling US Navy Douglas R4D is seen parked on the Douglas ramp; this was the period when the armed forces were using "day-glo" red as an anti-collision finish.

Electra N171PS flew with PSA from November 1959 until August 1968, and was bought back in August 1975, for service to Lake Tahoe. Sold to Evergreen International as N5539 in 1979, it was last reported in use by Channel Express Air Services as G-CEXS in April 1992.

Caravelle III F-WJAM was delivered to General Electric as N420GE, Santa Maria, on July 18, 1960. It was converted to a Series VII, with GE CJ-805-23C aft-fan engines, five months later. The twin-jet went to Air France as a Series III, F-BLKF, on July 30, 1963. After migrating to the Central African Republic as TL-FCA in 1982, it was stored in 1989.

By 1960, Sud-Aviation had sold more than 60 Caravelles, and being very anxious to place more of the type in the United States, entered into discussions with Douglas, for them to act as the sales agent for the type in major portions of the world. Douglas announced in 1960 that it had entered into such an agreement. As a result, a Caravelle III was sold to the General Electric Company, to have new aft-fan CJ-805s fitted in place of the Rolls-Royce Avons. The aircraft is seen on the ramp at LAX, prior to the engine change, in dual Douglas–Sud-Aviation markings, named *Santa Maria*. Failure to ensure sales of the type to US airlines canceled the agreement, but subsequently led to the development of the Douglas DC-9.

Douglas held a press conference for the Caravelle and then flew writers on an hour's flight out of LAX; the author was onboard. Seen during the takeoff are the new American Airlines and TWA maintenance facilities, just west of the short North-South runway. Construction of the two additional runways north of the terminal area can also be seen, with 24 Left almost complete. Despite the short length of the North-South runway, it was used occasionally by big jet airliners.

Returning to LAX in the Caravelle, we see Aviation Boulevard and the old terminal area to the left. In front of the old TWA hanger are a number of piston-engine aircraft. Aviation Boulevard had the Santa Fe railroad double-tracks running adjacent to it on the west side, with a spur across the highway into an industrial area east of the airport. This rail-line caused considerable congestion, preventing Century Boulevard traffic from getting into the airport whenever a train ran through. The problem was solved later by building a fly-over for the tracks. Note the new entrance way to Runway 25 Right, which allowed the jets to make wider turns onto the runway; remember the landing gear bogy problem mentioned earlier?

The old and the new were very mixed together in 1960, and here an American Airlines DC-6B makes its stately way into the new maze being developed at LAX. That year was really the last one for any number of piston types in service with the major airlines, partly because the major introduction of the big jets coincided with a downturn in airline traffic.

American took delivery of DC-6B N90757 on May 18, 1951, naming it Flagship Oklahoma, *and later* Flagship Oklahoma City. *It was sold to Hawaiian Airlines on April 9, 1964, and flown for two years before being stored. Yemen Airways bought the aircraft as 4W-ABD in 1968, stored it in 1971, then broke it up.*

N808TW was handed over to TWA on May 18, 1960. Withdrawn from use and stored at Kansas City, Missouri, in January 1974. It was bought by American Jet Industries as N804AJ four years later. In 1989, the 880 was used to replicate an earlier Boeing 737 fire accident, then scrapped.

In January 1961, TWA took delivery of the first of its second jet type, the Convair 880, and conducted a pre-inaugural flight out of LAX on January 10. The author rode on this and found the type an exciting one to travel in. Many speed records were broken by the 880s, but they proved to be heavy on fuel and maintenance, and only lasted until June 1974 in TWA service, primarily due to the oil embargo of 1973 (more on that later). Pilots liked the 880, but the cabin crews disliked its small galleys and the initial steep climb-out delayed start of onboard service, giving less time on the shorter routes it flew.

June 22–25, 1961, were the preview days for the new LAX "Jet Age" terminal, and many interesting aircraft were on display. The oldest was the Douglas M-2 mailplane, delivered to Western Air Express (forerunner of Western Airlines) in April 1926, and flown between Los Angeles and Salt Lake City. Note the new, vastly taller airport control tower in the left background.

This Douglas M-4 mailplane was one of nine bought by Western Air Express, who put the original M-2 type into service on April 17, 1926, between Los Angeles and Salt Lake City, Utah. It was re-acquired by Western Airlines for publicity and display purposes, and is shown here registered as the first M-2, NC150. The M-4 was actually NC1475, and was flown across the United States in 1977 for display in the National Air & Space Museum in Washington, DC.

Western took delivery of N7142C on May 17, 1961, converting it to a convertible passenger/freighter model in December 1968. The Electra spent most of the next ten years in South America, then went to Hawaii for two years, as N342HA. Passing through the mainland with Zantop International from 1985, it now resides with Fred Olsen Air Transport in Norway, registered LN-FON.

Western Airlines also displayed a rather more modern aircraft in the form of the Lockheed 188A Electra. It carries the striking Indian-head color scheme, also used on their DC-6Bs. Unfortunately, another Electra had crashed only three months earlier; that and the introduction of the jets, stopped any further orders, the total production falling far short of the projected 400 aircraft break-even point. However, after the structural problems leading to the crashes had been corrected, the type became a very smooth aircraft to travel in, as the author found on later Western flights.

The FAA displayed a very rare type, in the form of a civil Boeing 717-148. The 717 was the equivalent of the military KC-135, and its large forward cargo door is seen open here. The aircraft had only been delivered to the FAA a month earlier, originally bearing a USAF serial number.

N98 was a civil USAF Boeing 717-148, KC-135 (registered 59-1481), delivered to the FAA on May 20, 1960.

United received DC-8 52 N8036U on June 8, 1961. It went to Air New Zealand in November 1970, and ended up with Evergreen International as N804EV. After four years in storage at Marana, Arizona, it was broken up, in January 1984.

United Air Lines displayed one of its new Douglas DC-8-52s, with the later Pratt & Whitney JT3D-1 turbofan engines. These gave the DC-8 true intercontinental range in addition to making it considerably quieter.

One of Pan American's longer-range Boeing 707-331s is seen taxiing past construction still in progress at the new airport. This Boeing, in contrast to the United DC-8-52, is powered by the original JT4A turbojet engines.

A Series 331 originally destined for TWA, N703PA was instead delivered to Pan Am on December 30, 1959, as Clipper Dashaway. *It went to Air Manila International as PI-C7073 on December 12. 1973, finally being stored at Manila, Philippines, in February 1982.*

N94131 was Western's first 720-047B, delivered on April 7, 1961. Sold to AVIANCA as HK-723 on September 24, 1969, it was damaged beyond repair while landing at Mexico City, Mexico, on August 16, 1976.

A Western Boeing 720-047B gets away smartly on Runway 25 Right. The 720 was the "odd man out" in Boeing's commercial aircraft numbering system, due to United Air Lines. The carrier had delayed ordering the Boeings types, due to using Douglas DC-8s, and did not want to order aircraft with an earlier Boeing type number, which would highlight its tardiness. Boeing was only too happy to oblige with a higher number for such a major customer (however, it was never done again).

AIRLINERS AT LAX

Mexicana's fifth de Havilland Comet 4C taxiing out for takeoff in spring of 1961 displays the very ornate Aztec calendar on the tail. These were the only Comets to serve LAX commercially, but they remained in service for some 10 years. The type also was the only remainder of Britain's pioneering pure jet, which entered service in 1952.

Originally registered G-ARBB, XA-NAT, Golden Knight joined Mexicana on November 29, 1960. After being withdrawn from use and stored at Mexico City ten years later, it was bought as N777WA by Westernair of Albuquerque, but apparently not delivered to its new owner. The Comet was derelict at Mexico City, circa 1980.

N57204, Continental's fourth 720-024B, was delivered July 9, 1962. It later went to Ethiopian Airlines as ET-AFB on September 16, 1974. Boeing bought the aircraft in November 1985, for use in its KC-135E re-engining. Stored at Davis-Monthan AFB, it has since been broken up.

Continental's 720B displays a different kind of gold tail as it awaits clearance to turn onto the active runway. The 720 was considered to be the "hot rod" of the early Boeing family, being shorter, lighter, and having a different wing. Less fuel was also carried, and the type had a considerably better takeoff and cruise performance than the 707 series.

American Airlines bought 25 of the 720s after United's original purchase, but because they already had the 707 in service, called them "707 Jet Flagships." This name was actually painted on the aircraft. American's first ten were delivered with JT3C turbojet engines, while the next 15 arrived with JT3D turbofan power plants, and were designated as 720Bs. The first ten were later returned to Boeing for engine retrofitting. Note the North American sign for the X-15, visible below the jet's main gear.

N7544A was handed over to American on April 10, 1961. Sold to Middle East Airlines on December 31, 1970, as OD-AFN, it went to Omega Air, Eire, as EL-AKD in January 1991, and is now out of service, stored at Shannon.

N8811E was delivered September 9, 1961. Boeing acquired Delta's Convair jet fleet at the end of 1973, in trade for 727 tri-jets. Following freighter conversion, Fleet Number 911 went through several users before being withdrawn from service and stored at Caracas, Venezuela, in 1986.

Delta put the Convair 880 into service a year prior to this picture of its eleventh aircraft on approach to LAX on a "grungy" September 1962 morning, the local coastal cloud not yet having "burnt off." Delta's mostly white paint scheme made the aircraft look very clean and attractive.

TWA's striking red "arrow" fuselage stripe on the Convair 880 made a vivid contrast to Delta's color scheme, even in the "grunge." The type provided fast service between major cities within the United States, but was heavy on fuel consumption. This ensured its withdrawal when the Arab oil-embargo hit in 1973; most were withdrawn from use early in 1974.

After 13 years of service with TWA, N828TW was retired in April 1974 and parked at Kansas City, Missouri. It was bought by American Jet Industries in 1978 as N815AJ and moved to Mojave, California, for storage. Now owned by Torco Oil Company, the 880 remains in the California desert, still wearing basic TWA colors.

Mexicana bought this Comet 4C on January 14, 1960, as XA-NAS, and used it until December 1970. Also sold to Westernair of Albuquerque, in June 1974, it passed through the hands of several owners, and after being stored at Chicago-O'Hare from late 1976, was abandoned in 1979. Sadly, the Comet was broken up on-site in 1993, after an unsuccessful attempt to save it for preservation.

Also poking its way through the "grunge" was Mexicana Guest's fourth Comet 4C (The airline was continually having to get new financing to survive, hence the name change). This view clearly shows up the type's buried engines and very large flaps. There was a very heated debate in engineering circles for years, regarding jet engine placement (buried in the wing or in underwing pods), but there is no doubt that it would be impossible to "bury" the latest 90,000-pound thrust engines in any high-speed wing.

United's fifth 720-022, with the original JT3C turbojet engines, is seen on finals on a much brighter October day. Since taking delivery of its first jet aircraft, United went through many different paint schemes, as will be seen in later pictures.

Delivered on May 25, 1960, N7205U flew only for United. It was stored in 1973, sold in March 1976, and broken up at Minneapolis/St. Paul, Minnesota, in December 1976.

N8807E was acquired by Delta on August 7, 1960. On December 20, 1972, its vertical stabilizer was snapped off during a collision with a North Central DC-9-31 at Chicago's O'Hare International Airport. The 880 was considered beyond economical repair and stripped for spare parts, then scrapped the following summer.

Here is Delta's seventh 880 about to turn onto Runway 25 Right, with the original control tower still in position at the right. Although portions of the new airport facility had been opened some 16 months prior to the taking of this picture, they were then chiefly west of Sepulveda Boulevard, and it was a long time before all of the original facilities west of Aviation Boulevard were replaced by new air freight buildings.

A National Airlines DC-8-21 climbs out of LAX. This aircraft was the first of its fleet and is seen in the original color scheme. The sound suppressers are still in the rear position, but the flaps and landing gear are in the up positions.

National's N6571C joined the carrier on February 7, 1960; it was named Kathleen *in 1972. After having been sold in June 1973, the DC-8 went through many owners before being broken up at Islip, New York, in 1984.*

The first Electra built, N174PS flew in the certification test program as N1881, then was refurbished and sold to PSA. From October 22, 1968, Holiday Airlines operated it as N974HA. Withdrawn from service and stored, the prototype 188 was, sadly, scrapped at Oakland, California, in 1975.

PSA's six Electras were always coming and going at LAX, and one is on finals here. Behind the airline name on the fuselage there is the later "Super Electra Jet" statement. This was an attempt to eradicate the reputation of the original, unmodified Electra and also use the new in-word "jet" rather than the correct "turboprop." However, it wasn't too long before PSA began to acquire a replacement fleet of Boeing 727s.

By spring 1963, the jets had taken over most of the major airline routes, and the earlier large piston-engine types had all but disappeared. Many of the DC-7s were scrapped, but American Airlines had several of its DC-7Bs converted to freighter aircraft. One is seen on approach to LAX.

N347AA was accepted by American in September 1957, converted to a freighter five years later, then sold to West Coast Automotive in 1964. It flew for several cargo carriers before being acquired by TBM, Inc. as a spares source. The aircraft remains on the US registry as N756Z.

JA8002 joined JAL on July 29, 1960. American Jet Industries bought it in June 1974, registered as N420AJ. The DC-8 was converted to a freighter in December 1979. Stored at Mojave, California, from April 1982, it was broken up eight years later.

Japan Air Lines opted to use the DC-8-32 on the long Tokyo–Honolulu–LAX run and here is its second example, named *Nikko*, taxiing in after landing. These DC-8s served for almost 14 years with the airline.

Scandinavian Airlines System's first DC-8-32, *Dan Viking*, is seen climbing out after takeoff. Landing gears are all fully up, but flaps are still in the takeoff position. SAS used these DC-8s on its long, over-the-pole flights, and as a result, takeoff runs could be extremely long on hot days. This one is seen in May 1963; three months later, the author saw one aircraft takeoff that did not rotate until the main gears had reached the stripes at the far end of the runway.

OY-KTA was with SAS from March 31, 1960, until September 1971. Passing through many owners, it was converted to a DC-8-33 freighter in October, 1978, and ended up flying cargo out of Peru as OB-T1316 in 1988. It was damaged beyond repair after overrunning the runway at Iquitos on August 10, 1989.

Clipper Viking was delivered to Pan Am as N723PA on October 27, 1959. Sold to GATX/Boothe Aircraft Company in May 1970, it had a checkered history, ending up in Ankara, Turkey, in 1980. Now owned by the Turkish Authorities as TC-JCF, it is still in storage at Ankara.

Pan American was among the first to use the 707s on Pacific routes, and one of its 707-321s is seen taking off, with the new airport control tower and Theme Building behind. This takeoff was very much shorter than the SAS one referred to previously; it is seen while the landing gear is still retracting. Water-injection-caused engine smoke is much in evidence.

This is Mexicana's de Havilland Comet 4C XA-NAT (seen also on page 62), awaiting clearance to turn onto the new 24 Left runway, behind a Pan American 707. Note that by spring 1963 the name of the airline had reverted to just Mexicana, without the previous "Guest" suffix.

Mainliner Kauai, a DC-8-21, was delivered as N8027U on May 25, 1960. It was sold to Sun Land Airlines in November 1978, and was converted to a freighter the following year. Last leased in 1984, it was broken up at Miami, Florida, in February 1986.

United Air Lines fleet of DC-8s continued in heavy use on their cross-country routes, and here is its 27th on finals. By spring 1963, the drag problems with the original DC-8 wing had been rectified, resulting in very intense competition with American Airlines 707s on transcontinental routes.

Not far behind the United DC-8 in the previous picture, was a very clean Flying Tiger Line Lockheed 1049H Constellation. This was the freighter version, and Flying Tigers used it for almost nine years on the gradually growing air freight business. They utilized the new cargo facilities at the east end of LAX, on the north side of Runway 25 Right.

Flying Tigers bought ten 1049H freighters, and N6912C, Fleet Number 802, was the second, delivered on March 25, 1957. Leased out in 1966, the aircraft was finally sold to Blue Bell, Inc., on February 1969, but not operated by that company. Downair of Canada later acquired the Connie, as CF-BDB. Following another change of hands, the aircraft was broken up for spares in 1979.

N811TW spent its entire flying career with TWA, from February 2, 1961, until retirement on November 2, 1972. Following storage, it was scrapped at Kansas City, Missouri, in May 1977.

By the spring of 1963, TWA's Convair 880s were well-established, and here a very clean one is seen taxiing for takeoff. The new terminals at the airport were very much farther apart than the old facilities, and this required longer taxiing to reach the allocated runway. Most domestic flights continued to use Runways 25 Left and Right, while the longer international flights usually used the new, stronger, longer Runways 24 Left and Right. Located west of Sepulveda Boulevard, they were not subject to the weight limitations imposed by the runway bridge over that highway.

As mentioned earlier, PSA's six Electras were very frequent users of LAX. Here one is seen on finals to Runway 24 Right. Note that all reference to "Electra" had been removed from the airframe as, by summer 1964, most travelers preferred pure jets.

PSA bought N181H, originally ordered by Capital Airlines, on May 17, 1963, and applied a new registration: N376PS. It was sold to Universal Airlines May 24, 1968, and converted to a freighter later that year. The Electra passed through several users, ending up with Zantop International Airlines in June 1982.

F-27A N2771R was delivered to Pacific on April 2, 1959, and became part of the merged Air West fleet in March 1968, then Hughes Airwest on April 1, 1970. It went through a series of leases in France, Sweden, and Bolivia, being finally stored at Dinard, France, in 1990.

Fairchild's F-27s were also still in frequent use by Pacific Air Lines in summer 1964, and one is seen on finals, in the latest color scheme. Among the smaller local airlines in California, it had managed to survive, like Bonanza and West Coast Airlines, but its future was in doubt due to the increasing expense of new aircraft. However, it was the cost of fuel that would prove to be the determining factor in their future.

Continental Airlines continued to provide very popular service (some people thought the best) out of LAX, but like many of the airlines at that time, found it difficult to finance upgrades to its aircraft. In summer of 1964 we see one of the original 707-124s on finals. Unlike the American Airlines 707-123s converted to turbofans in 1961, it still has its original JT3C turbojet engines.

N74612 was the fifth 707 delivered to Continental, on March 17, 1960. TWA bought the aircraft in December 1967, and then sold it to Israel Aircraft Industries four years later. It was finally bought by Israel Airports Authority in September 1983, to be used for ground training at Tel Aviv.

N106A was delivered to Eastern Air Lines as a model 649 on June 5, 1947, and converted to a 749 in 1950. Sold to California Airmotive in January 1961, it was leased to Trans California and several other carriers before ending up in Dominica as HI-129 in January 1967. The Constellation was removed from service in 1974, and broken up in March 1977.

By the summer of 1964, the once large fleets of Lockheed Constellations on passenger flights had been withdrawn from use. They were relegated to freight service and startup airlines. One of the latter is a Trans California 749, used mainly on trips from LAX to Lake Havasu, a growing resort center on the Colorado River.

A very uncommon visitor to LAX in July 1964 was this Finnish Kar-Air charter line Douglas DC-6B. Bought only a month earlier from Northwest Orient Airlines, it became famous by being converted to a swing-tail freighter configuration, four years later.

This Douglas DC-6B was bought by Northwest Orient as N577 on October 5, 1957, and sold to the Finnish carrier Kar-Air O/Y on June 12, 1964. They carried out a swing-tail cargo conversion in April 1968. Returned to the US in June 1982 as N867TA, it was sold it to Northern Air Cargo, in May 1986.

N721PA Clipper Splendid was Pan Am's eighth of the type, delivered October 19, 1959. Withdrawn from use in September 1970 and stored at Wichita, Kansas, it went to Dan-Air London on January 1, 1971, as G-AYSL. KIVU Cargo bought the 707 January 10, 1983, and broke it up for spares.

Pan American also continued to use its 707-321s with their original turbojet engines; one is seen landing in July 1964. The lower ventral fin, not an original feature on the early 707s, was added retroactively. This was caused by the British Air Registration Board requiring additional fin area to be added to BOAC's 707-436 versions before certifying them as safe for service. Extra height was eventually added to all 707 (and USAF KC-135) vertical tails to meet the Dutch roll stability requirements; as a result the lower ventral fins were not fitted to most later 707s (they had also prevented full nose-up rotation of the aircraft for takeoff on shorter runways).

American Airlines changed its color scheme in 1964. One of its turbofan-powered 707-123Bs is seen on finals in July 1965, with the new livery (other airline employees sometimes called this the "naked-nose" or "lightning bolt" scheme).

As with many other 707s in American's fleet, N7551A was sold to the Boeing Military Airplane Company on April 14, 1983, and used in the USAF KC-135E program. The remains were stored and later broken up at Davis-Monthan AFB, Arizona.

N37522 was named Mainliner Wyoming when it first arrived from Douglas, on July 11, 1947. Following an uneventful career with United, it was sold off to Mars Aviation in 1968, and scrapped at Ontario, California.

Even as late as July 1965, United Air Lines was still using some of its venerable DC-6s. Here is N37522 on finals, with a very clean finish, despite having been in service for 18 years.

At the beginning of 1966, Bonanza took delivery of three new Douglas DC-9-11 twin-jets; at the end of the year the third one is seen taxiing past the new TWA LAX terminal. The tremendous demand for the DC-9 caught Douglas by surprise, and together with personnel and engine shortages due to the growth of the Vietnam War, caused such production and financial problems that the company was forced to merge with McDonnell in April 1967.

N947L, the 37th DC-9 built, was delivered to the airline July 1, 1966. When Bonanza became part of the newly formed Air West, the twin-jet was returned to Douglas and converted to a DC-9-14. After several assignments, it was sold in 1991 to Estrellas del Aire of Mexico, as XA-RSQ, named Tres Estrellas de Oro.

N946L joined Bonanza on January 17, 1966. As with its sister aircraft, the DC-9 went back to Douglas, on May 16, 1969. Following conversion to a DC-9-14, it flew for Finnair. Republic Airlines bought the aircraft on December 6, 1985, registered N930RC. As Fleet Number 9140, it remains active with successor Northwest Airlines.

The second Bonanza DC-9-11 is seen against a very attractive sky, in spring 1968. Note that the name "Funjet" had been added behind the airline name on the fuselage. The DC-9 became extremely popular with the airlines, greatly outselling the Sud Caravelle and the BAC 1-11. A key reason was the ability to carry larger amounts of freight in the belly, providing extra revenue. The aircraft was quiet and smooth, after an initial engine/fuselage vibration was cured. These early versions also suffered from very hard landings (like Navy carrier aircraft), but a change to the landing gear oleos solved that problem.

We mentioned earlier the financial problems the smaller local carriers faced. On April 18, 1968, Bonanza, Pacific, and West Coast all merged to form Air West. This is the first Boeing 727-193 originally delivered to Pacific Air Lines in 1967, about to leave the ramp. It is painted in the short-lived first color scheme of the new Air West. Note the ramp agent sprinting to get out of the coming blast of the three jets.

N898PC left Air West less than a year after the merger, being leased to Braniff. It was then sold to Burma Airways as XY-ADR, in 1970. After several changes of ownership, it was finally leased out under Hong Kong registry, as VR-CBV, in September 1993.

Delivered to National Airlines October 23, 1962, N278C acquired the name Carolyn *in 1972. It was leased at the end of that year to Air Jamaica, and bought by them December 15, 1973, as 6Y-JGF. Capitol Air purchased the DC-8 on July 1, 1983, re-registered N921CL, and stored it at Smyrna, Tennessee, where it was eventually relegated to firefighting practice.*

By the summer of 1968, National Airlines DC-8-51s carried the new "Sun King" insignia on the tail. This later version of the DC-8 was powered by JT3D-1 turbofans and the new engine nacelles no longer had the characteristic rear-moving silencers. The increased power allowed the aircraft to carry larger payloads, and reduced noise levels considerably. This is the third of the airlines' original DC-8-51 order, delivered in 1961.

Western's third Boeing 707-347C is seen taxiing for takeoff on August 3, 1968, only six days after delivery to the airline. It is in the long-lasting color scheme, with the Indian head at the front of the fuselage stripe. Western had relied on the Boeing 720B up to then, but traffic growth caused the addition of five 707s. In 1970, five more were ordered but not added to the fleet.

N1503W was acquired by Western on July 29, 1968. Sold to TAAG Angola Airlines as D2-TOM in August 1980, it was destroyed by fire at Luanda, Angola, on October 10, 1988.

Trans-Canada took delivery of DC-8-41 CF-TJB, Fleet Number 802, on May 25, 1960. Renamed Air Canada June 1, 1964, the airline sold the DC-8 to Transavalair in June 1977. It was withdrawn from use and broken up at Sion, Switzerland.

Air Canada flew DC-8-41s into LAX from Montreal via Toronto, and its second example is seen taxiing in the center of the airport. These aircraft were powered by Rolls-Royce Conway engines, rather than JT3s, due to Commonwealth preferences not requiring tariffs on the engines when imported into Canada. The unusual black anti-glare panel on top of the nose gave them a slightly menacing appearance.

In the late 1960s and early 1970s, British Caledonian Airways flew many charter flights from LAX to Gatwick, England, using Boeing 707s that enjoyed very high load factors. This was not only due to low fares, but also to the superb onboard service by their Scottish tartan-clad stewardesses (the author enjoyed several of these flights). Caledonian had recently merged British United Airlines into its system, hence the Caledonian/BUA markings, with the rampant Caledonian Scottish lion on the tail. The aircraft is at the south side terminal, next to Imperial Highway.

Originally ordered by British Eagle, Boeing 707-365C was first leased in 1967, then bought in 1968, by Airlift International as N737AL. It was sold to Caledonian Airways as G-ATZC, County of Stirling, *on July 7, 1970, then went to Transbrasil as PT-TCP, on February 4, 1986. The Boeing was damaged beyond repair on takeoff from Manaus, Brazil, on November 26, 1992.*

Fairchild F-27A N154L was delivered to Bonanza Air Lines April 5, 1962, being re-registered as N754L in 1964. It served with successor company Air West, then Hughes Airwest, until February 1974. After flying in the South Pacific and Mexico, it was stored in 1987 at McCarran Field, Las Vegas, Nevada.

Only two years after Air West was formed, it was bought by the legendary Howard Hughes, and became Hughes Airwest, on April 1, 1970. A startling new paint scheme was applied to the aircraft, as seen on this Fairchild F-27A taxiing to the north side runways in October 1973. The purple and yellow hues were the parent Hughes Aircraft house colors, and came from those owner Howard Hughes used on his prewar racing aircraft (now on display in the National Air & Space Museum). At least the livery was very visible and distinctive. (Two years after Hughes' death in 1976, his parent company, Summa Corp., sold the carrier to Republic Airlines, effective October 1, 1980.)

After the end of the Vietnam War, a great many changes were going through the airline industry. A drop in the war-induced traffic was aggravated considerably by the Arab Oil Cartel embargo in early 1973, which caused all fuel prices to sky-rocket (gasoline cost less than 50 cents a gallon before then). Airlines began to adopt new color schemes in an attempt to stand out at airports. Here is a Western Boeing 737-247, taxiing for Runway 24 Left. It carries a striking new fuselage stripe with a giant "W" at the front, and the name "Western" in large black letters all the way up the tail. Often called the "pygmy jet" at LAX, the 737 eventually passed Douglas' sales of the DC-9/MD-80 series to become the world's best selling jet transport.

Western's N4524W, acquired March 17, 1969, was sold to Aloha Airlines on March 13, 1976. Its subsequent history is very involved, and includes at least 18 airlines all over the world. Now with International Air Leases, Inc., it is based in Mexico with Magnicharters, as XA-STB.

XA-TAC was delivered to Mexicana on November 9, 1970. Later it was sold to Aeron Aviation Corporation, then leased to Turkish airlines Bogazici and THY, between 1988 and 1990. Colombian airline AVIANCA flew the tri-jet for two years, until it was returned to the United States as N434ZV, in September 1992. It was purchased by JetAir Trading in May 1994 and stored at Mobile, Alabama.

By 1975, a large number of airlines had acquired fleets of Boeing 727s, especially the more economical -200 version. This is Mexicana 727-264, *Mexicali,* on finals, showing the large flap area. Early versions of the 727-100 had an even lower flap maximum-down position, but several accidents were caused by pilots under-estimating the very high drag in that condition. The solution was elimination of the last flap down position; the type subsequently gained a very good reputation.

Western's new color scheme had a great visual impact on its larger Boeing 720-047Bs, as seen in this view of one on final approach. Painting the entire fuselage white, rather than just the upper portion, added to the overall effect. One certainly knew which airline flew the type.

After 15 years with Western, N93148 was bought by International Lease Finance Corp. on May 18, 1978, and sent to Kenya Airways as 5Y-BBX. It was stored at Nairobi, Kenya, in January 1992, and scrapped later the same year.

Convair 580 N73301 was bought from Frontier Airlines by Sierra Pacific on November 18, 1974. Used by the airline until stored at Marana, Arizona, in the 1980s, it was leased to Resort Commuter Airlines in the late-1980s, and has since been converted to a freighter.

A new small local operator operating into LAX in the mid-1970s was Sierra Pacific, which operated between LAX and the Mammoth ski area. One of its Convair 580s is seen on approach in May 1975. The attractive paint scheme was typical of the period.

By the mid-1970s, most of the major long-haul airlines were flying Boeing 747s into LAX, and here is a close view of the nose of a United 747-122 just prior to touchdown. It is wearing the "Four-Star" livery, which was a slight variation of the airline's first 747 colors. As we will see, United adopted a simpler corporate identity a short time later. Note how the main landing gear bogies trail, while the two center ones remain parallel to the aircraft.

N4719U joined United June 26, 1971. It was named Friendship Japan *in 1983 and remains in active service.*

Delta accepted N9898 on November 22, 1970. Returned to Boeing when the airline discontinued 747 service, it was leased, then sold to China Airlines as B-1860. Guinness Peat Aviation purchased the jumbo jet several years later and had it converted to a freighter. Passing through several companies, the aircraft is now owned by Evergreen International, registered as N479EV.

Delta was also flying 747s into LAX from Dallas and Atlanta; here we see one "reaching" for the runway. The sheer bulk of the 747 is very apparent in this view, and also the great height of the cockpit above the runway at touchdown. The type has been a big money-maker for Boeing, and it will remain the largest commercial aircraft in service into the next millennia.

This is another United 747-122 just about to touch down, but shown in the striking new color scheme referred to previously. It is obviously a repaint, as the aircraft registration is immediately prior to the one in the earlier picture. We had mentioned earlier that there was a drastic drop in passenger traffic in the early 1970s. The author remembers one United night flight from Kennedy Airport, New York, to LAX when the cabin crew out-numbered the passengers in the rear cabin.

Named Thomas F. Gleed, N4718U was delivered new to United in 1971 and remained active in the Friendly Skies for 26 years.

N554PS was acquired by PSA March 15, 1974. Pan American bought it in November 1984, applying the registration N375PA and name Clipper Flying Cloud. *Sold December 5, 1991, it has been stored and operated on short-term leases. The last known owner, from February 1995, was Jet Lease Finance.*

Meanwhile, PSA did not intend to be left behind on the West Coast, and had exchanged its Electras for Boeing 727s. In this unusual view, we see one of the more than 30 model 214s that PSA ordered (though not all of them were taken up by the airline), on finals for 25 Right, with a giant Northwest Orient Airlines 747 in the background, also on finals, but lined up for Runway 24 Right. The striking new color scheme on the PSA 727, with a "smile" line on its nose, contrasts with the more sedate red, white and blue scheme on the 747. This was the period when the PSA stewardesses wore their famous red "hot pants" outfits; being mostly long-legged California blondes, they garnered a great deal of attention.

In March 1968, Lockheed announced 144 firm and option orders for its new L-1011 TriStar airliner. TWA had ordered 33 plus 11 options, and put the type into service on its longer routes. Like the competing DC-10, this type proved popular with the public, but its introduction coincided with a major downturn in traffic. Here is TWA's ninth TriStar on finals to Runway 24 Right.

Delivered to TWA as Fleet Number 11009, N31009 arrived on May 16, 1973. It was returned to the leasing company in 1992 and then operated by Air Operations Europe as HR-AMC. Withdrawn from service at the end of 1994, the TriStar was scrapped nine months later, at Bournemouth, England.

N113AA flew with American from April 1972 until its retirement in July 1993. Stored at Amarillo, Texas, it has since been sold to Federal Express Corp.

American Airlines introduced the wide-body Douglas DC-10-10 in August 1971, flying between Chicago and LAX, just 11 days before United put the type into service between Washington, D.C., and LAX. This view shows the most unusual feature of the DC-10; its center tail-mounted engine. This power plant position and the area-ruled cockpit windshield made the DC-10 a very quiet and smooth aircraft to fly in, the author having made many flights to Europe, South America, and the South Pacific in various versions. This example served for more than 21 years with the airline.

Eastern Air Lines began service into LAX on September 23, 1969, with nonstop flights from Atlanta. Here one of their Boeing 727-25Cs is just about to touch down, marked with the colorful US Bicentennial seal of 1976 on the forward fuselage. Eastern called the 727s "Whisperjets," which they most certainly were not!

Acquired new by Eastern in November 1967, N8164G was operated in both passenger and freight configurations until it was sold to Federal Express on August 31, 1982. Now carrying cargo only, the 727 has been re-registered N128FE and is named Stuart Shawn.

CP Air took delivery of CF-CUR on March 11, 1971. It went back to Boeing on March 17, 1977. After service in Bolivia, it was used by the Revlon Corp., then sold to the Baker Corp. in 1986.

In Canada, CP Air (owned by Canadian Pacific Railways) competed with the then-government-owned Air Canada, and eventually obtained rights to fly into LAX from Vancouver. One of its early 727-17s reflects a new color scheme with the famous "bite" tail marking. Boeing bought the aircraft back only 10 months after this picture was taken, CP Air having purchased two more economical 727-200 replacements.

Pan American was the first to order the Boeing 747. *Clipper Pacific Trader* is shown taxiing out for takeoff on Runway 24 Left. Delivered to Pan Am in April 1971, it still carries the small size airline name on the forward fuselage.

Delivered to Pan American as N652PA on April 25, 1971, Clipper Pacific Trader was renamed Clipper Mermaid in 1980. It was removed from service and stored at Orlando, Florida, in April 1991, then bought by Polaris Leasing Corp. and converted to a freighter. Leased out several times, it was re-registered as N652SJ, then as N832FT by the current lessor, in 1995.

N532PA, first flown by Pan Am, migrated to United in February 1986, acquiring the registration N142UA. It was retired at the end of 1994 and is now stored at Las Vegas, Nevada.

Pan Am began flying the first of ten Boeing 747SP-21s in early 1976 to handle very long nonstop flights, including those from LAX to Japan. Here we see *Clipper Constitution*, only two months after delivery. Somewhat of an expedient derivation of the 747, the type had a shortened fuselage, requiring a much taller vertical tail to maintain stability. One new problem pointed out on its long flights was the need for improved cabin air-conditioning systems, with provisions to relieve the extreme dryness at the high altitudes flown. Pan American's fleet was sold ten years later to United.

Texas International started as a local airline known as Trans-Texas Airways. By 1976, it had begun leasing DC-9s from TWA. A company DC-9-14 is seen getting away smartly on Runway 25 Right. It is hard to believe that Texas International could end up acquiring Continental Airlines, but this happened in October 1982, just over six years after this picture was taken.

Originally bought by West Coast Airlines in 1966, N9103 went by merger to Air West. Purchased by Texas International on November 8, 1975, and named City of Laredo, *it crashed at Baton Rouge, Louisiana, on March 17, 1980.*

N31030 joined TWA on August 27, 1975. Converted to L-1011-100 standards in February 1978, its fleet number was changed from 11030 to 31030. The TriStar was sold and leased back in December 1989, and has since been retired from service with Trans World.

A close-up view of the nose of TWA's thirtieth TriStar on finals to Runway 24 Right.

Not to forget the local commuter airlines flying into LAX in this period, we see a de Havilland (Canada) DHC-6 Twin Otter of Golden West Airlines. These Twin Otters provided an excellent method of avoiding the long drive to LAX from major outlying areas, such as Orange County, Long Beach, and Ontario, and gave passengers a very good view of the ever-expanding greater Los Angeles basin (though from a very crammed cabin).

N63118, a DHC-6-200, was originally delivered on April 11, 1968, to Aero Commuter, which merged with Golden West Airlines in July 1969. It was bought by Missionary Aviation Fellowship in August 1983, re-registered as P2-MFR.

N712DA, Fleet Number 712, joined Delta on September 18, 1974, and remains active with its original owner.

Delta also bought the Lockheed L-1011 and ended up with 28 of the basic -1 model. This is its twelfth aircraft, about to take off on 24 Left. Delta's attractive color scheme blended in very well on the large vertical tail and integral intake duct, tending to reduce the visual size of the latter. This engine mounting was one of the major differences between the TriStar and the DC-10. Some thought it more elegant than the DC-10 installation, but it led to some power loss due to duct drag, and created a strange rumbling noise in the rear cabin. The author flew several trips in the TriStar, and found it to be very much on a par with its competitor. However, the bitter fight between McDonnell Douglas and Lockheed for orders finally led the latter to pull out of the commercial airliner business, and prevented Douglas from ever reaching its break-even point on the DC-10.

Air Canada also bought six of the -1 version of the TriStar, benefiting from the lack of excise tariffs on its British RB.211 engines. Here we see their distinctive paint scheme with the stylized Canadian Maple Leaf on the vertical tail. Note that the airline still used the black anti-glare panel on the top of the nose. This aircraft is also about to begin its takeoff run, with the leading edge slats out and trailing edge flaps in the first down position.

Air Canada took delivery of TriStar C-FTNK on March 23, 1974. Designated as Fleet Number 511, it was converted to a L-1011-100 in January 1977. Leased to Air Lanka as 4R-TNK on December 29, 1981, the TriStar returned to Air Canada with its original registration late in the following year. Retired from service and stored at Marana, Arizona, in November 1990, it was bought and leased out to Royal Airlines in 1993.

N1819U was handed over to United on April 12, 1974, and flew with the carrier until its accident at Sioux City.

This United DC-10-10 is getting ready to roll for points east. N1819U was United's nineteenth, and became well known as the one that crashed while attempting an emergency landing at Sioux City, Iowa, on July 19, 1989. This accident was the result of an engine disc failure which caused an almost total loss of hydraulic fluid in the aircraft control systems. The crew did an incredible job in maintaining a measure of control almost to the touchdown, saving many lives in the process. Note that the aircraft is still finished in the earlier "Four-star" livery.

TWA had the distinction of owning the longest-lasting fleet of Boeing 707s in major airline use. Here is one of its 707-131Bs about to take off, complete with an attractive new paint scheme. Note how small the 707 now looks, compared with the newer 747s, L-1011s, and DC-10s, only 17 years after its debut. This clearly illustrates the tremendous growth in airline traffic during that time period, despite the inevitable ups and downs.

N6722, Fleet Number 6722, operated for TWA from April 28, 1966, until May 19, 1982, when Boeing bought it back for use in the USAF KC-135E program. The 707 was stored at Davis-Monthan AFB, Arizona, and later scrapped.

Delivered new to National on June 16, 1975, Fleet Number 83 became Pan American Clipper Celestial Empire when the airlines merged in January 1980. American Airlines acquired it four years later; the registration N142AA was applied. The tri-jet was removed from service and stored at Marana, Arizona, in May 1995.

Not to be outdone by all those TriStars, here is one of National Airlines' DC-10-30 aircraft taxiing out for takeoff. This view illustrates a little-known feature of the long-range version of the DC-10. The center main landing gear could be left in the retracted position if the aircraft weight did not require its use; in this view it is clearly in that position. Note the girl's name "Tammy" painted just behind the cockpit, a feature applied to all of the airline's jets. At the time, this aircraft was only 18 months old.

Japan had grown to a major industrial power during the 1960s and 1970s, and the Japan Air Lines 747 illustrates the expansion of Asian commerce. This view emphasizes how far forward the passenger cabin extends on the 747, due to the unique third-level flight deck (which came from the nose-opening proposed for the Boeing C-5 entry).

JA8106, a 747-246B, was delivered on April 14, 1971. Sold to Hissin Lease October 31, 1994, and leased back, it continues to fly for JAL.

American accepted N854AA on June 22, 1976. Aircraft Leasing Inc. purchased it in January 1995 for freighter conversion and lease to Kitty Hawk Air Cargo.

Boeing 727s carried a great deal of the passenger traffic in the US at this time, and here is a beautifully clean American Airlines version on finals. American remained steadfast to the natural metal finish on its aircraft, being virtually the only major airline to retain this basic livery.

Photo Index

Page	Date	Airline	Aircraft Type	Registration	Manufacturer's Serial Number
6	12/1/56	TWA	DC-3	N86544	11689
7	8/5/56	TWA	049 Const.	N90823	2085
8	8/5/56	Delta	DC-7	N4874C	44264
9	1/27/57	United	DC-6B	N37568	44081
10	12/1/56	TWA	1049G Const.	N7105C	4586
11	1/27/57	United	C-54	N30043	10487
12	8/18/56		Bristol Britannia	G-ANBJ	12911
13	8/18/56		Bristol Britannia	G-ANBJ	12911
14	3/10/57	United	DC-7	N6329C	45143
15	3/17/57	Continental	DC-6B	N90961	44689
16	12/1/56	TWA	C-54B-1-DC	N34577	10541
17	3/23/57	Western	CV-240	N8408H	47
18	3/23/57	Western	DC-6B	N93116	45067
19	3/27/57	Western	C-54B-15-DO	N86573	18383
20	6/6/57		Sud Caravelle I	F-BHHI	02
21	6/6/57		Sud Caravelle I	F-BHHI	02
22	3/30/57	Austral	C-46	LV-FSA	30323
23	7/14/57	TWA	1649A Starliner	N7303C	1004
24	8/4/57	TWA	1649A Starliner	N7311C	1013
25	8/4/57	TWA	1649A Starliner	N7320C	1023
26	8/4/57	Catalina Pacific	DC-3	N55L	26675
27	8/4/57	United	DC-7	N6334C	45148
28	2/23/58	Southwest	M-202/DC-3	N93056/N63105	9146/20213
29	8/2/58	United	CV-340	N73112	16
30	8/2/58	Continental	DC-7B	N8211H	45194
31	8/2/58	Bonanza	DC-3	N498	1903
32	8/2/58	Pan Am	DC-7C	N739PA	44881
33	5/9/59	Pacific	DC-3	N54370	19220
34	5/9/59	Bonanza	F27A	N147L	38
35	5/9/59	TWA	049 Const.	N90831	1970
36	6/1/59	American	707-123	N7507A	17634
37	5/9/59	TWA	707-131	N732TW	17659
38	8/22/59	Delta	DC-7	N4878C	44682
39	8/22/59	Continental	707-124	N70775	17611
40	1/10/60	United	DC-8-11	N8012U	45289
41	1/10/60	Pacific	Martin 202	N93047	9133
42	1/10/60	Western	DC-6B	N93131	45536
43	1/10/60	Western	188A Electra	N7136C	1070
44	1/10/60	Continental	Viscount 812	N244V	357
45	1/10/60	TWA	707-131	N739TW	17666
46	1/17/60	PSA	188C Electra	N172PS	1109
47	4/60	United	DC-7		
48	5/60	Mexicana	DC-6	XA-JOT	43213
49	5/60	Mexicana	DC-6	XA-JOT	43213
50	5/60	PSA	188C Electra	N171PS	1091
51	8/31/60		Sud Caravelle III	N420GE	42
52	8/31/60	(Press flight)			
53	8/31/60	(Press flight)			
54	9/21/60	American	DC-6B	N90757	43269
55	1/12/61	TWA	CV-880	N808TW	22-00-10
56	6/25/61	Western	Douglas M-2	NC150	244
57	6/25/61	Western	188A Electra	N7142C	1128
58	6/25/61	FAA	717-148	N98	17969
59	6/25/61	United	DC-8-52	N8036U	45303
60	6/25/61	Pan Am	707-331	N703PA	17680
61	6/25/61	Western	720-047B	N93141	18061
62	6/25/61	Mexicana	Comet 4C	XA-NAT	6443
63	7/62	Continental	720-024B	N57204	18419
64	9/62	American	720-023B	N7544A	18030
65	9/62	Delta	CV-880	N8811E	22-00-50
66	9/62	TWA	CV-880	N828TW	22-00-35
67	9/62	Mexicana	Comet 4C	XA-NAS	6425
68	10/8/62	United	720-022	N7205U	17911
69	10/62	Delta	CV-880	N8807E	30
70	3/63	National	DC-8-21	N6571C	45391
71	12/1/63	PSA	188C Electra	N174PS	1001
72	3/3/63	American	DC-7B (F)	N347AA	45400
73	5/63	JAL	DC-8-32	JA8002	45419
74	5/63	SAS	DC-8-32	OY-KTA	45384
75	3/63	Pan Am	707-321	N723PA	17601
76	4/10/63	Mexicana	Comet 4C	XA-NAT	6443
77	4/63	United	DC-8-21	N8027U	45296
78	4/63	Flying Tigers	1049H Const.	N6912C	4809
79	5/63	TWA	CV-880	N811TW	22-00-14
80	7/64	PSA	188C Electra	N376PS	1133
81	7/64	Pacific	F27A	N2771R	43
82	7/64	Continental	707-124	N74612	18012
83	7/64	Trans Calif.	749 Const.	N106A	2523
84	7/64	Kar Air	DC-6B	OH-KDA	45202
85	7/64	Pan Am	707-321	N721PA	17599
86	7/65	American	707-123B	N7551A	18883
87	7/65	United	DC-6	N37522	43011
88	12/66	Bonanza	DC-9-11	N947L	45730
89	4/68	Bonanza	DC-9-11	N946L	45729
90	8/68	Air West	727-193	N898PC	19620
91	8/3/68	National	DC-8-51	N278C	45643
92	8/3/68	Western	707-347C	N1503W	19965
93	8/3/68	Air Canada	DC-8-41	CF-TJB	45443
94	6/71	Br. Caledonian	707-365C	G-ATZC	19416
95	10/73	Hughes Airwest	F27A	N754L	91
96	10/73	Western	737-247	N4524W	20128
97	5/75	Mexicana	727-264	XA-TAC	20434
98	5/75	Western	720-047B	N93148	18588
99	5/75	Sierra Pacific	CV-580	N73301	80
100	5/75	United	747-122	N4719U	19880
101	5/75	Delta	747-132	N9898	19898
102	5/75	United	747-122	N4718U	19879
103	5/75	PSA	727-214	N554PS	20875
104	5/75	TWA	L-1011-1	N31009	1029
105	5/76	American	DC-10-10	N113AA	46513
106	5/76	Eastern	727-25C	N8164G	19720
107	5/76	CP Air	727-17	CF-CUR	20512
108	5/76	Pan Am	747-121	N652PA	20347
109	5/76	Pan Am	747SP-21	N532PA	21024
110	12/76	Texas Int'l	DC-9-14	N9103	45796
111	12/76	TWA	L-1011-1	N31030	1111
112	12/76	Golden West	DHC-6-200	N63118	118
113	12/76	Delta	L-1011-1	N712DA	1088
114	12/76	Air Canada	L-1011-1	C-FTNK	1069
115	12/76	United	DC-10-10	N1819U	46618
116	12/76	TWA	707-131B	N6722	18988
117	12/76	National	DC-10-30	N83NA	46714
118	5/76	JAL	747-246B	JA8106	19825
119	12/76	American	727-223	N854AA	20995